The Old Direction of Heaven

The Old
Direction of Heaven

POEMS BY JENNIFER ROSE

TRUMAN STATE UNIVERSITY PRESS
NEW ODYSSEY SERIES

Printed in the United States of America.

Cover art: Edna Amit (Lilly Bobasch)

Library of Congress
Cataloging-in-Publication Data

Rose, Jennifer.
The old direction of heaven: poems / Jennifer Rose.
p.   cm.
ISBN 0-943549-23-X (pbk. : alk. paper)
I. Title

The paper in this publication meets or exceeds the
minimum requirements of the American National
Standard—Permanence of Paper for Printed Library
Materials, ANSI Z39.48 (1984).

for my grandmother, Blanche Seelmann,
and for Holly Hickler

in memory of Beatrice Hawley

A HUMAN LIFE, I think, should be well rooted in some spot of a native land, where it may get the love of tender kinship for the face of earth, for the labors men go forth to, for the sounds and accents that haunt it, for whatever will give that early home a familiar unmistakable difference amidst the future widening of knowledge: a spot where the definiteness of early memories may be inwrought with affection, and kindly acquaintance with all neighbors, even to the dogs and donkeys, may spread not by sentimental effort and reflection, but as a sweet habit of the blood. At five years old, mortals are not prepared to be citizens of the world, to be stimulated by abstract nouns, to soar above preference into impartiality; and that prejudice in favor of milk with which we blindly begin, is a type of the way body and soul must get nourished at least for a time. The best introduction to astronomy is to think of the nightly heavens as a little lot of stars belonging to one's own homestead.

—GEORGE ELIOT, *Daniel Deronda*

Contents

Acknowledgments

Some of these poems originally appeared in *The Nation* ("Eastham Sonnets," "Horoscope," and "Lipik Postcard"); *Paris Review* ("Strawberries" and "On Losing the Sense of Smell"); *Ploughshares* ("At Dachau with a German Lover" and "Cologne's Cathedral"); *Agni Review* ("Britons Leaving France," "Mostar Postcard," and "Lines Written during an Autumn Invasion"); *Partisan Review* ("The Italian Rose Garden"); *Verse* ("Solstice"); *Boston Review* ("First Frost in the Suburbs"); *Chelsea* ("A Morning Walk"); *Compost* ("Saratoga Journal Entries" and "Catechizing the Dandelion"); and *The 1989 Grolier Poetry Prize Anthology* ("Letter from Paradise"). "At Dachau with a German Lover" also appeared in the anthology *Naming the Waves* (Virago/Crossing Press). "Midnight Swim" is included in Louis Karchin's "Songs of Distance and Light" (1988).

Many thanks to the Massachusetts Artists Foundation, the Waltham Arts Lottery, Virginia Center for the Creative Arts, the Mary Anderson Center and the Corporation of Yaddo for their kind and generous support, and to Sydney and Norman Abend for time in Eastham. Special thanks to Holly and Fred Hickler, Martin Edmunds, Carol Moldaw, Randi Schalet, Alice McMahon, and Edward Rose for many years of enthusiasm and good advice.

At Dachau with a German Lover

for Nico Reichelt

I won't go with you to Munich's planetarium
though I have always loved a wandering moon.
I cannot bear to bless a German heaven.

Dachau. The sign appears, colloquial
amidst the traffic; the radio sputters
Stau —or is it *heil?*

Everything continues in this language!
Every chimney rises with a grudge.
The *Arbeit* gate swings slowly on its hinge.

This is the first time I feel at home
in your country, in this museum.
Elsewhere, the Nazis are innocuous —your neighbors.

You examine by yourself their careful orders.
Every road to death is neatly chartered.
You're horrified —not just by deaths of strangers,

but by the language, which killed them
before the gas or gunners;
your language, words you might have uttered.

Tell me what it says, this chart of stars:
Which color is my destiny of fire —
yellow, for the language of my prayers?

red, for the fury of my cares?
or pink, my crime of twin desires?
This is my planetarium, these pinned-on stars!

1

You won't go with me to Dachau's crematorium
though the ovens are cold, the fumes are gone,
and you're too young to have fired them.

The gas chamber, familiar as a dream
but smaller, is open at both ends.
I walk, without knees, without lungs, the brief

avenue. This is it. The vault, the safe
where they escaped, scratching
in the old direction of heaven.

At night our room is dark, the bed, a ditch.
The moon grows big in its hutch as we watch,
wide awake, tense. Your father was a soldier,

your mother, a Hitler youth who quit.
I'm a Jew. I'd be dead if we were older.
What shocked me most as we first slowed

to stop was not Dachau's walls or weight,
but its shamelessness at its own sight.
Had no one seen the guards guarding their flawed height,

the smoke drifting off, signaling in desperate code?
This is what I can't forget:
how public it was, how close to the road.

Cologne's Cathedral

for Gary Lee and Bob Harris

Cologne's cathedral rises in a steep geyser
of rust, stopped fountain of ore, backwards torrent
of brimstone and extinguished prayers.

Its volcanic thrust ruptures the cement
outside the rebuilt Bahnhof where streamlined
trains hurry the tourist to her dank ascent

inside this umber chunk of Gothic mind.
For six centuries of mornings, faithful masons rose
to hoist the bricks of this pyramid, by hand.

Burnt effigy of God, it still glows,
darkly, towering over the heads of curious Germans
wandering in its ash of ornate shadows.

And yet—each buttress droops like a singed wing,
earthbound, too heavy or weary, as if
those centuries of witness were crippling.

Its rusting voice calls in a guttural clef,
hoarsely invoking the industrial
hours, as it called to the mason, scaling the roof,

attaching at last the beacon finial
and aiming its light through history's thick dust.
This may be God, this chain of touch, not the menial

zest preached from the apse with such pious lust.
Here, where the spire shrinks at its wilted timberline
and the crockets' burrs are fastened thickest,

here, above the Rhine's busy railroad span
and the trestle-necks of cranes, God is not burdensome.
Cologne's real prayers are said in its construction.

The Dom itself rests on a millennium
of Roman manufacturing ruins. The glass
Colonia's kilns produced revives its flames.

And every night at eight, the guidebook promises,
blond floodlights bathe this bishop's house of sand.
Its bleached pigment like a tan fire blazes,

reaching toward heaven with a long, nicotine-stained hand.

Britons Leaving France

Boulogne-sur-Mer

Boulogne's iron rooftops close like umbrellas
over the houses, half open, half shut
like its citizens on their shallow balconies.
Terraces of newer edifices lean out

but no one is home there.
The cathedral spreads out its tower
like a blessing, with a pontifical air
and a stony confidence blessings are needed here.

The Union Jack billows on the British ship's mast
like the chest of a pre-war soldier.
Local boys run along the long pier,
clutching themselves and diving at last,

a desperate gesture as the boat glides past and past and past.
On deck and shore wavers gather
to salute each other.
Though they are strangers and this the local sport,

they wave as if their effort
would heal the continent,
its crevices of evidence of war,
or this gray rupture of famous water.

The waves return to France's shore
like small animals afraid of water.
Gulls push the ship out of port,
past the boys running, past the beach

blank of bathers where one small dog cavorts,
oblivious, past the red family
strolled out to the red lighthouse,
anchored by infants and a Sunday pram,

past the new moon impaled on the cross,
past the language of bread, the delicate,
untranslatable vocables
of love filling the rooms on the hill. . . .

Suddenly the gulls flap but do not call,
the children grow bored with their perch
by the lifeboats. Dusk and water claim
the horizon at the same moment.

A man unfolds his *Daily Mirror*
and sends his daughter after tea.
A line forms at the Bureau de Change.
Already there is an announcement

of a lost child. Everything is in English now —
all pretense of French a quaint formality.
All passengers adopt the ruddier nationality
though they were French a moment ago.

Amphibian, European,
they finger their passports, check the familiar
watermark of their queen,
only for a moment estranged.

It is a group effort, home.
They must all agree.
The gulls lag behind,
rags on a kite string.

When the child is found,
it is a British victory.

The Italian Rose Garden

1

Today they are adjusting the Roman fountains
and stocking new goldfish in the basins.
The opulence suggests they could adjust
the roses' blooms. And Rome was even gaudier
than these sandblasted-white fountains intimate,
pale imitations of the resplendent rose,
tourniquets of the rose. A white Ceres
too short for a goddess clutches wheat to her breast
on a pedestal. She looks out over
the hacked-back bushes, peering through her cataracts.
Blind to the roses' fury, she does not suspect
their intentions yet: arming all civilians.

2

Because I know this is an Italian garden,
I see Mussolini speaking from the stone terrace.
Behind him, the pergola seems a corridor
less meant for roses than for long queues
of deportees. Wires strung between the columns
contain them in rows. Then suddenly,
the statues look amateur, too crude,
a ruse: cement bricks instead of soap cakes.
And are the birds metaphors
for planes or angels? Is the sun God
or a searchlight? The roses—some sort of ambush?
Manifestos kill the pink Rugosa.

3

Late afternoon soft-focuses the garden
with pollen. Now it is a photograph.
With its gardeners and tourists weeded out
one cannot guess the year or circumstance.
The drone of the highway beyond the fence
could be a thunderstorm or planes landing
at an aerodrome. Shadows trickle over
the green terraces. This could be peace.
Why must I feel like Nero fiddling when I praise
the fountain and the rose? The rose is neither
a fire engine nor the fire itself.
And neither are the birds' songs sirens.

Mostar Postcard

Bosnia–Hercegovina, Yugoslavia, 1982

Soldiers on holiday fill the outdoor café.
A young man has just jumped off the old bridge
for their money, on his chest a tattoo
of their leader, a hero hired by heroes.
These men are the same age as I am
but theirs is the age of the native-born,
drawn on their handsome faces with a dark pen.
Peasants, mechanics, farmers — I would not dare
to call them brothers. One winks. Some stare.
I pretend to look at the river.
How war makes borders glamorous,
our time brief and historic!
But this is not war, just the face of war;
not love, just faces of lovers.
Love could launch Mostar's minarets —
love, or war — though language stops us here:
dawdling at the river's edge
where history whispers in each soldier's ear.

Letter from Orahovica

July 19, 1941, Croatia

Earlier that day he was with us as we posed in the field at dawn for a photograph, as we smoked and shaved by the river, as we burned every house in the village. I don't recall his step but his aim was always accurate. When the captain called us, he was neither first nor last in line.

Click, click. Hans's camera framed the partisans we'd caught, propped up against the massive haystack, joined at their clammy hands like cut-outs, identical in their blindfolds. We waited until Hans stepped back. The captain gave his terse command, off-hand as a rich man ordering his car. (This was to let the peasants know how unimportant they were.) Josef — Schultz — seemed not to have heard the captain's quick *Bereit!* and stood dreaming with a shell-shocked look. *Ready,* the captain called again, this time bitter, impatient, not for the prisoners' benefit, but as if he knew what was coming. Josef stared but did not move. The air smelled of smoke and hay and sweat. The captain questioned. Josef put down his gun. The captain made threats. Josef began to undress — helmet, cartridge belt, all the signs. Then he was through and walked toward the haystack. After four steps, he turned back, dropped his dog-tag into his helmet. Not one of us had dropped his gun from his shoulder. Then Josef took his place in the other line.

The next man in the other line, though blindfolded and unable to understand German, took Josef's hand, making a last friend. Hans's clicking sounded like a bloodless gun. Josef stared and did not say a word. The captain made a sign and then a low sound. The peasants leaped forward, obedient as dolls. Josef still stood, friendless; none of us had shot him. Hans clicked. One of the peasants gurgled. The captain quickly ordered fire and Josef was dead. Then the captain took his pistol and fired two more shots into Josef's head.

Lipik Postcard

Croatia, Yugoslavia, 1982

Lipik's graveyard is lush as tropics,
blushing a hundred wet blooms,
the village women tending them.

A vendor sent me to this garden,
directing me in perfect Deutsch
and promising that I could reach

the grave of Alfred Miller,
my great-grandfather,
son of the village innkeeper,

who died hiding, running
with partisans, his yellow star on,
shot near the war's end by Nazis (Croatian).

Across from the vendor's kiosk
was the mustard-colored inn,
its name worn thin —so thin

one more rain might have ruined it.
But genealogy resists
the picturesque: no grave exists

for Alfred here, there's nothing
I can make bloom like these women can,
pumping and watering all afternoon.

Only one grave here has been forgotten:
a German grave, overgrown with weeds
and buried behind three feet

of dried vines. One evergreen
is the cenotaph I claim.
The tulips' goblets now toast Alfred's name.

Even Here

Gulls bobble like Buddhas in the bay.
Hermit crabs drag their small hotels.
Robins sight-read symphonies.
Serenity in the clef of a shell.

Grass banks this house like a vast steppe,
one skeletal sailboat beached in its green
waves. Spikes of goldenrod poke up
and thistles guard the dune's dominion.

Legions of waves, like columns of soldiers
parading Red Square, arrive forever,
their medals ornate as Titanic chandeliers,
their shore leave never quite over.

Horseshoe crabs' helmets rust among shards
of civilian shells. Eel grass holds the beachhead
as no army ever could. To be buried
here, whales crash of their own accord.

High tide piles up the slipper shells' sabots,
the seaweed-hair. Images of that war,
even here—though how can one suppose
to know it, among these metaphors?

Black-shelled snails stud their reversed firmament,
slow comets. Then the ocean covers their heaven
with its fog. A dozen trawlers crawl the blank horizon.
Above Billingsgate's Atlantis, permanent

waves. Surf's gray shingles topped that island's houses
while fog's lost film recorded the streets' farewell.
Waves search the bay each night like empty hearses
but the island still lies buried where it fell.

Lines Written during an Autumn Invasion

A pheasant pulls the pin of its sudden croak.
The woods have been napalmed with frost.
Purple loosestrife crusts, sanguine maroon,
dyeing the river with a tempera wake.

Overhead, planes cast avian shadows.
Willows —young women prematurely gray —
await the crow's telegram. Still the rolling stone
of river portages its phosphorescent moss

and squirrels ignore the wind's grave diagnosis.
Still blue jays laugh and thrashers sing
and the milkweed gives its science class,
in spite of everything.

First Frost in the Suburbs

The marigolds' barn-color chips and fades.
Their shrivelled buds hang down, knots of dried wood.

A slate squirrel, thin in the wrong season, drags
its twine tail like a spent fuse. Stacks

of trees smoke with the flesh of leaves.
The petunias' popped-balloon skin flaps empty sleeves.

The frozen lawns—square fields of captured hay;
a mutt slinks across them like a practiced refugee.

Now every chalky nimbus holds its bank
of snow. Cold bands of sparrows walk the plank

of sky. A small orchestra of birds warms up at dawn—
frantic and lyric as Auschwitz's musicians.

Saratoga Journal Entries

1

Winter here is brown, dark green and cadmium.
The rare cry of an invisible bird is
the only sound in the forest.
After a coup, starkness of the new regime.
All week I have felt like a tourist here
where everything else wears snow like a fashionable fur
and the footprints are no bigger than fists.
The lakes have been smoothed by a calender
of snow. Conifers make a lattice
over the shrouded furniture.
Soon guttural voices will thaw the cathedral
and the sumac's red scars will heal.
But first the magnolia's buds will be killed
by frost and a shadow delay the whippoorwill.

2

Icicles spill from the roofs in great cataracts.
Snow and seasons clog the fountains.
A sculptor curates the garden with chairs
though the tourists' footprints are mingled with bears'
and the statues hibernate in wooden shacks.
The flower beds are rolled-up carpets,
the rose bushes stacked chairs. Winter has no clocks
in its rooms, just shrouded furniture.
Summer is a cheap souvenir regret
can save, old calendar still turned to June
in this cold room in the Adirondacks.
My heart is tired of facts. But winter's white dialect
is too scientific to explain spring's appeal
and talk of love gets too rhetorical.

3

In the frozen mansion one's lungs turn
into radiators which hardly heat the room.
Ghosts leap from the mouth in cauls of breath.
The shrouded furniture holds them like urns
of summer. In thawed mirrors they groom
the past. In their minds the lawns are green
under the white footprints of trillium
and birds drink from the defrosted fountains.
They are like horses dreaming of races.
Blank hours in the cold stall of winter
do not cancel their paces, these white epiphytes
I make. They live forever in the stalled calendar
of memory, like the souls of statues.
No cataract of snow will cover their faces.

4

Downtown's pavilions are deserted oracles,
the pilot lights of the mineral springs
shut off like old gods. The Grecian urns
are sown with snow. I'm the only tourist.
The war dead suffer another burial.
Their kiosk is a piece of shrouded furniture
no one would buy. Even men who returned
had their names listed here and now those are lists
of more dead. Memory is the only souvenir
we have of them—and that is melted snow
soon enough. While seventy civilian springs
have passed, these heroes' lives were kept in frozen
escrow. Statues renew their stony curfew.
The groundhog's shadow predicts a longer winter.

5

The clock ticks, a souvenir from when time
mattered. The fountains will not be mirrors
for months. Trillium is still a ghost.
Wind wins every race at the track. I don't
want to fill the snow's blank calendar
with anything—not even words. I've come
for memory, not conversation, a tourist
less interested in statues than in shrouded
furniture. If I could preserve the soft fossils
of bodies on sheets like those footprints of stars,
I would. I want evidence love is immortal.
But bodies disappear like the angels
one makes in snow—no matter how well-guarded—
and so does love, its fountains, its thawed footprints.

Horoscope

for D.H.

The moon floats above us, Orion's
lost discus, hurled in another season. Polaris
beckons us, Pegasus rides towards heaven.
Our kiss casts a penumbra on each face.
How am I to know what love is?
Venus is a distant muse, her yellow
meridians swathed in shadow, bouquets
of nebulae her only guarantee.
Gravity, the comets' alibi,
provides the heart with its rhetoric.
In the planetarium of debris,
a phosphorus wand tattoos the zodiac.
All winter, above the dead trees, Orion flees
the Scorpion and seeks the Pleiades.

Regret

for S.D.

Cartoon balloons of breath escape pedestrians' mouths
as I drive home. Spring in Massachusetts
is a poem about snow and broken hearts.
Even streetlights are snowflakes, fat moths

on the windshield, without my glasses.
Myopia yields its metaphors
most beautifully at night. Affairs
are what we spoke about this evening. My guess is

I'll never have one and you'll have more —
though I regret this. Jazz on the radio
requites the rain's romantic notions. The snow
has changed to sloppy kisses. I admire

your mistresses, you know — the fact of them.
Freedom is my favorite abstraction, next to love;
regret, the thing I'm most afraid of.
Still, one can't make love to a theme

and the Girl of My Dreams is already over.
Tonight, while winter is having its last fling,
yet still wearing its white dress, I long
to kiss someone new, tell her I love her

with the white amnesia of a storm
about morning. I long to regret
the thing I've done — instead of haven't. You requite
your loves; I go home and write yet another poem.

Fog

The factory's smokestack minarets call
me to prayer. Fog muffles the muezzin's wail

or would if there were one. Sodium lamps rust
the fog till it stings the night's tongue like a first taste

of beer. I wish you were here
instead of across the river. Over here

the fog's more like champagne. The lilacs' grapes
have made it, I guess, or Bacchus perhaps.

The neighbors' tulips clink their goblets of it.
Drunk on it, I might admit—what?

Something about love, no doubt; but I'm not.
Cars sibilate down velvet streets that separate

our houses. A commuter train unzips its route.
The campanile's eclipsed clock clangs eight,

nine, ten; all night its landlocked foghorn warns
me what not to do—not even once.

Now rhododendrons reappear in sips.
Dawn gulps the fog as if at last it were your lips.

The Kiss

The night heron perched in one tree since noon
turned back his clock yet another time zone.

I felt like moaning. Had passion been a crayon
the foliage would have doubled its wattage.

The city's broken church bells gonged in my head,
their lost hours at last resurrected.

Poplars' still-green Valentines were airmailed
by the wind, a thousand leaves with one name on them.

An osprey flew over the river
where gulls usually swirl, their perpetual mobile

stilled by the world's sudden vertigo.
The dam spilled its suds like the secret we'd just told

each other and a train thudded into the station.
Sparrows then fluttered a timely ovation.

Sonnet to a Married Man

The bodies of winged ants lie crumpled like clichés
of thrown-out drafts, their fertile buzzing stalled
by poison. On my head, the burden of those canceled
generations. Love has been like this

for me — these crash landings of small bodies,
these tight skeins of something barely unraveled.
As I sit among these things I have killed
I think of your kisses and all those losses.

A female ant — three black beads of mercury —
drags her dead partner across the gray floor,
graceful reversal of the caveman's violence.

Your kisses, their guilty innocence,
make me sorry for your fidelity
and more. Today I saw anthills everywhere.

Strawberries

These strawberries announce desire like roosters,
their reveille lasting till well past noon.
No mouth domesticates their primal sirens,
no apron can carry enough sweet souvenirs.

Our mouths are primed for kisses like these each morning,
your beard soft as the straw between these rows
of berries. My hands search the shadows.
Everywhere buds are turning to tongues.

Solstice

The sun loiters above me in its tropic,
weighting summer's scales in love's favor
with the gold coins of this longest day's hours.
The bees too dawdle through their sticky ecliptic.

Pinks' staccato laughter tickles the bed
and the deep red voice of the geranium.
The ear of the nibbled peony suggests a worm.
Far from winter, a cricket splurges; the wood

lily aims all the anthers in its quiver.
This garden is vicarious with pollen,
the thistle's exile less wandering
than desperation. The sun hovers,

perfect lover, an immortelle. I browse
among these bleeding hearts, still unrequited.
Seasons of passion like this are not suited
to harvest: the rose burns out; the orchids bruise.

Home after Three Weeks Away

The sky has faded like an old brochure
for summer. Clouds derail above Bear Hill.
Spring is over; petunias trumpet the solstice.

The sour chimes of an ice cream truck repeat
their theme, a jacked-up lullaby.
Silence here is only contrapuntal.

Roses glut their vines, late June's speeded-up
assembly line. Now every night our legs mesh
skillfully — gears in the factory

that makes the roses. But I love you
like the gardener, not the bee.
Yes, exile provided its altitude

but now that is done. Headlines divide
the mind again. Traffic outside elides
the O's of tires. Some truth eludes me.

Homesick pigeons return to the yard.
The windowbox explodes with latent buds. Still
the concrete neighborhood hoards its best words.

On Losing the Sense of Smell

In my arms, lilacs like a former lover—
no heartbeat, no elevator.
The garden pantomimes its tactics. Now

the ocean arrives too soon: flora
the only clue before the roar of blue.
I forget what it is to be far away

and still know I am near. The wind is
an empty boat. Overboard, bouquets
of memory. Nostalgia cannot translate

mnemonics. Summer goes black and white.

Eastham Sonnets

1

Fishermen drag in the tide. Their reels sound
like gulls. Clam diggers forsake their spit of muck
for the cold sand where you and I walk,
barefoot, against the flag-flapping wind.
Do they guess this is a "romantic weekend"
for us or do they think two women cannot ache
that way? You will not take my hand in public,
afraid of the fishermen's reprimand.
I think this fear in you is what disturbs
me most, this desire for us to act like fugitives.
The hermit crab teaches you a lesson
not meant for us. Love, why must you listen?
All around us lies the wreckage of lives,
shells abandoned like poisoned suburbs.

2

Enameled crabs are strewn in brittle
constellations on the beach. The bleached clams
are dented moons. Their astrology is subtle;
every planet makes a tidal claim.
I wonder as we walk which one you will choose.
Gulls' prints decorate the wet flat in hieroglyphs
that look like birds themselves, fifth-grade arrows
chasing the clouds. We follow their path as if
at the end we, too, might take flight. Thirteen
summers ago I rescued a gull here
which died on the way to the Audubon station.
I made love with a boy on these dunes. The shore
is a palimpsest of loss; nothing resists synthesis.
But the sea has no pity for the sacrifice.

3

Eel grass rattles in the dunes, silver,
overexposed, a drained monochrome.
It gathers the sand in its hairy weir
with a gray grasp. My claim to you is the same
tentative clasping. Wellfleet's foghorn booms
an elegy over the filled-in bay.
Dusk shuts its blue shutters against the storm.
Tonight perhaps the tide will be high
enough to wash the beached slum of sea trash away
from the house, to set free the loosened buoy
of memory. You are my only buoy now.
The sea imposes its restless curfew
on us. Gulls patrol the shore. The beach house
harbors our nervous exile. Waves lap at the grass.

4

Light the color of white wine pours out
of the neighbors' window. Ringing the bay
are smaller lights, sweet bells. The target
ship disappears into indigo.
It has been anchored in the bay ever
since I can remember, a porous reef
of shelled metal. One summer, still during the war,
my uncle rowed me to it in his skiff
when the bomber squads weren't practicing.
Light flickered like minnows in its cavities
while the sea completed the bombers' instructions.
War and the sea have settled so many destinies.
I am thinking now of the soldiers rescued at
Dunkirk and how rarely love sends its boats out like that.

5

Morning. The sand is snow again, a white carpet
that needs vacuuming. The tide has waned;
soon the clam-diggers will return with their buckets
and hip-high boots. The sea sounds moccasinned.
Gulls reel above you as you try to find
some suitable memento to take home
among the broken necklaces of shells and strands
of seaweed jumbled like cheerleaders' pom-poms.
Now you bring me the casing of a razor clam,
the sea's slim switchblade, though you fail
to notice its blunt symbolism.
O how I envy the grip of the barnacle,
impervious to the tide's commute;
how I envy the handsome fisherman his net.

Poem

Sinkers of light have been plunked in the dark ink
of the river, white and sodium pink
trajectories. If the path of memory
could be preserved, it would look like this
as it entered its black hole. The river is
both sleep and insomnia, like poetry.

Driving home alongside it, I begin
to notice its nuances, how the windows
distill and the streetlights are swallowed;
how Harvard's lit-up Parthenon glows,
a candelabra of those candles which do not blow out,
leaning like Narcissus to his own reflection.

And now I see luminaria
bobbing like those cast off each August
by the children of Hiroshima.
It is hard to believe I never noticed
till now these flickering lilies tapping the current,
funnels of light trailing their phosphorescent

roots. Suddenly, I find myself swerving
for a look at the river where it diverges
from the road, I find myself promising to learn
its terms and navigation of its rapids.
I think I will follow it all night, down dark roads
and through the soft yards of sleeping cottages,

past sluices and dams, marshes and streams,
past the last settlement of modest A-frames
bordering its banks. I will follow it
past everything I have known, I will even swim
in it if I dare disturb its white throats,
I will float among those flasks of echoes. Time

grows dim. Love and immortality
do not matter to me now, as long as there is this river
dilating its flowers, as long as it will accept one more
ingot of starlight, one more swan among its tears —
as long as I may return again to its shores,
nothing matters to me now except poetry.

A Morning Walk

Sweet Briar, Virginia

Sunday. No semis surge by.
I cross half a dozen finish lines
that spiders strung last night between the boxwoods.
Twee-twee-twee trills a bird never named by Adam.
Cool sun, blurry, a scoop of melting sherbet,
drips its light over the meadow.
Crickets ring softly, a neighbor's telephone,
but no one answers. The cicadas' alarms
have not gone off yet. A freight train
clacks metallically toward New Orleans.
Two cardinals, still dressed for evening, head home.
Steers graze down by the stream, auditioning
for Joseph's dream unknowingly. I envy them
their green reprieve, its blissful ignorance
and final usefulness. A vetch twists toward me
like the serpent's question mark. What have I
done to deserve Paradise, or exile
from it? Buzzards twirl, a silent mobile,
or an angels' carousel—without the choir.
Far below, a bunting's too-blue suit dazzles the pews
and goldenrod overpowders her nose—
as usual. The crow repeats an old sermon.
Still, every field full of insects is
a rapt congregation murmuring amens.
Their primitive worship humbles me,
like the sound of hymns healing
some ramshackle chapel's cripples and sinners.
O for a Sabbath not spent grappling with faith!
An arthritic cedar offers me
his mumbled benediction like a wizened deacon.

Dragonfly-seraphim hover around him,
agile as helicopters. At last God beckons
and I surrender, and then another poem begins.

Catechizing the Dandelion

Catechizing the dandelion, I kneel in the grass
though I lack the Catholic tourist's portable sign
of faith, that quick cross.

I have no way to worship at this white cathedral
detaching its prisms.
I have not memorized its Latin epithets.

For one franc, lights illuminate its transepts ten minutes.
Still I cannot decipher the frescoes of dust.
Still the icons are hidden.

But all afternoon, the wind has been proselytizing,
converting the dandelion, bead by bead—
Brittle reliquary! Dimmed nimbus!

Its tonsured head
now bows like mine, bent to appraise
the mysteries of faith. . . .

I think of the ruined church
that sits in the fields near Cuise-la-Motte.
Its bricks are scattered like the trepanned skull

of an anchorite left to rot.
But faith germinates in the burst pod of its skeleton
and tapers of yellow dandelions flicker near the altar.

At Mount St. Francis

Waterlilies grope the lake's green air
as hungrily as koi, which populate
a pensive pool nearby. Some friar
has made this garden shrine, no doubt,

including the St. Francis looped with plastic beads
like hoops around a prize at county fairs. One bird
imitates the sound effects of pinball games,
as lyrical as David in his way. A psalm

fragment (the twenty-third) displayed for us
to contemplate seems obsolete beside
the actual waters and pastures mowed
in yellow swirls by the T-shirted novice.

What made him sacrifice the world of touch?
Hibiscus blooms promiscuously
while the grapes say their green rosary.
Could I love anything as much as the church

of trees, its gospel sung by thrushes and some tone-
deaf chickadees? I cannot guess his motives.
He prays. I write. We're both alone. Soon an unseen
hand will light the waterlilies' votives.

Diaspora: Eden

for Myra Sklarew and Gardner McFall

Birds muscle through their blue aviary.
Trillium's dropped handkerchief triggers the hours.
A squirrel leaps through staffs of bracken,
playing the notes of some musical score.

Here fish are a species of angel,
treading reflections of heaven like planes,
none of them too literal for transformation.
Columbine is a red swan, bowing;

spikenard, the x-ray of an angel's lungs.
The vocables of birds defy transliteration
and rhubarb wags its dog-tongue
in a dead language—the missing link between

its exiles and this green ghetto.
Crows test the acoustics of paradise.
Someone invents the echo.
Who to thank for their blank language?

The applause of grass and trees has lasted centuries.

Letter from Paradise

Birds call, antiphonal, like cicadas
or jackhammers and the concrete's echoes.
A pendulum of shadows swings across
the lawn. The clouds have been cut out with jigsaws.
Copper dragonflies mate in flight; the noon sun is
an alchemist (though now the puzzle pieces
of clouds drift closer). A menthol breeze
dries the dew here each morning. Butterflies
flap their Texas-shape wings. Time has
no plus or minus; it just is.
(Owls hoot, regardless, expert witnesses.)
One night, as I walked among the pine trees —
and not even the stars could see this,
it was so dark —sap dripped from the branches
like oil from cooling engines or wax
from candles. Every hiss of grease
was a sigh. And now what I have been wondering is
whether or not all those clichés about paradise —
how insidious it is —are true. Crows
caw-caw, black kites stuck in trees.
Finches bathe in the fountains. Roses
climb the trellis where I dawdle on the terrace
and a squirrel plays. I have begun to count the days.
These vistas grow customary as the cries of jays.

Midnight Swim

The Big Dipper ladles out our portion
of heaven. The moon is generous —
candles for all the years of civilization.
This is paradise: ailanthus fringes

the pool; the moon beams like the countenance
of some deity. The poplar, by day
dappled with coins, now settles its accounts.
Newly-minted stars shine, more currency

to buy love with, and time. Privilege roams
the chrome lawns in the toga of night.
Rows of pines lean in, forming a stadium
around us with their quiet silhouettes.

They are a green as deep as our sadness
at leaving. The moon keeps their green from going black
the way poetry keeps sadness's nuances.
Lights from nearby houses shine through the thick

réseau of branches, modest moons. We discuss
how the fool who tried to capture the moon
in his rain barrel found in the morning it was
gone, his makeshift vault not lossproof. How even

the waxing moon has a wick which burns down.

Notes

At Dachau with a German Lover

The inscription above the main entrance gate to Dachau concentration camp reads *Arbeit Macht Frei*: "Work makes you free."

Identification patches were given to all prisoners according to their classification. For example, Jews were given yellow stars; political prisoners were assigned red triangles; homosexuals received pink triangles. Dual classification was also possible, e.g., Jewish political prisoners wore stars made from a red and a yellow triangle.

The Italian Rose Garden

The Nazis disguised their gas chambers as showers in order to calm their victims. Cement soap cakes were part of this deception.

Saratoga Journal Entries (1)

A *calender* is a paper-smoothing machine.

Eastham Sonnets (4)

During World War II, British soldiers surrounded at Dunkirk on the coast of France were rescued by a flotilla of small pleasure and fishing boats after Churchill called on private citizens to volunteer their aid.

About the Author

JENNIFER ROSE was born in Evanston, Illinois, in 1959 and has lived in Massachusetts since 1971. She has been a "Discovery"/*The Nation* winner and is the recipient of awards and fellowships from the National Endowment for the Arts, Massachusetts Cultural Council, Astraea Lesbian Writers Fund, and the Poetry Society of America, among others. Her poems have appeared in *Poetry, The Nation, Antioch Review, Ploughshares,* and elsewhere. This collection was a finalist for the T. S. Eliot Prize for Poetry. She works as a city planner specializing in downtown revitalization.

About the Artist

EDNA AMIT (Lilly Bobasch) was born on October 20, 1927, in Czechoslovakia. She was deported to Terezin on June 26, 1942. On October 4, 1944, she was deported from Terezin to Auschwitz. From there she was transferred to Freiberg and then to the concentration camp in Mauthausen, from which she was liberated in May 1945. She now lives in Israel, where she is retired from a career as a teacher of handicrafts. Her painting is part of a collection of children's drawings from Terezin preserved in the Jewish Museum in Prague.

Cover design by Tyler Schmitt
Truman State University designer

The poems and text are set in Optimum Roman 11/14

This book was printed and bound by
McNaughton & Gunn, Inc., Saline, Michigan